Acknowledgem

CW01664457

We would like to thank the following for tl... publication.

Transporting Somerset, Somerset County Council, for their involvement and support.

David Ainsworth for his care, diligence, professionalism and support in editing this book.

Taunton School for the use of their Minibuses (Picture).

South West Coaches for the use of their Minibus (Picture).

Open Government Licence

No Warranty

Contents

The Safe Operation and Driving of Minibuses

Introduction

This booklet is designed to give guidance to managers and those who are driving or planning to drive, minibuses either as a professional or casual driver.

As a driver, you have a legal responsibility for the safety of all of your passengers and other road users. Therefore it is essential that you fully understand your role and the 'duty of care' you have to others when driving. We recommend that all drivers read this booklet before undertaking their duties as a 'minibus driver'. It is also advised that drivers read the latest copy of the Highway Code which can be purchased from bookshops or downloaded from www.gov.uk

Legal Responsibilities before you Drive

Operators licence

Vehicles used for hire and reward within the UK generally require an Operator's Licence. The term '**Hire and Reward**' means any payment in cash or in kind by or on behalf of the passengers. The operation of vehicles for hire and reward will require an operators licence. Licence must be displayed in the lower near side of the windscreen and are colour coded dependent on the type of licence.

Permits

The use of a permit, or permits, allows the operator to use vehicles on a 'not for profit' basis and is valid for a maximum period of five years. Charges or contributions towards the cost may be taken but in most cases it will not cover the total cost of running the vehicle. Vehicles with fewer than nine seats can now be included in the use of permits. Examples of organisations which are eligible to operate vehicles under the permit or 'not for profit' scheme, include the following:

- **Religious**
- **Education**
- **Social or welfare**
- **Organisations which benefit the community**
- **Recreational**

However, there are different types of permit dependent on the organisation and how the vehicle is used. **Section 19 Permits** tend to cover the majority of journeys undertaken by most local authorities.

The authorities' Fleet Managers are usually responsible for the issuing of permits and should be contacted in the first instance with any questions relating to the use of permits.

Licence to Drive Motorised Vehicles

Most UK drivers now hold a two-part licence - a photo card licence and (at the time of writing) a second, paper part, of the licence. Some drivers may still hold the old-style paper licence.

It is worth noting that the government is currently issuing only the photo card licence. The types or categories of vehicle that a driver may drive are listed on the rear of the card licence. In order to establish if a driver has any restrictions or points on their licence, the DVLA has set up a new website www.gov.uk/view-driving-licence or, alternatively you can telephone the DVLA on 0300 790 6801 in order to obtain further details.

For most drivers, the photo card licence is valid for ten years. However, for professional drivers, the professional categories on the licence are generally only valid for a maximum of five years.

Should any driver suffer any medical problems which may affect their ability to drive, they must report this to the DVLA. Any failure to keep your driver's licence up-to-date can result in a maximum fine of £1,000.

Car licence obtained prior to 1st January 1997

If any driver has passed their car driving test 'B'category licence before 1st January 1997, in addition to their category 'B' licence to drive a car, they will also have the categories C1, D1 and E on their licence. This is known as 'grandfather rights' and enables them to drive a vehicle or a combination of vehicle and trailer in excess of 3.5 tonnes. Please note that restrictions will apply and drivers should check with the DVLA to ensure that they are not in breach of the regulations.

Those drivers who passed their test on or after 1st January 1997, face additional restrictions on what they are legally eligible to drive. (See table overleaf).

Those drivers who drive professionally for 'Hire and Reward' will also need to obtain a driver qualification card (known as a Driver CPC card).

Category	Descriptions
B	Drivers may drive a motor vehicle with manual transmission weighing up to 3,500 kg with no more than eight passenger seats plus the driver. The driver may tow a trailer of up to 750 kg. However, the combined weight must not exceed 3500 kg
D1	**Minibuses** with between 9 and 16 passenger seats,and a total vehicle length of not more than eight metres. with or without a trailer weighing no more than 750 kg.
D1 + E	**Minibuses** with between 9 and 16 passenger seats and eight metres in length with a trailer weighing over 750 kg. The weight of the vehicle and trailer cannot exceed 12,000 kg . The trailer must not be more than the maximum permissible towing weight for that vehicle.
C1	**Medium Sized Vehicles** between 3,500 kg and 7,500 kg with or without a trailer of 750 kg.
C1 + E	**Medium Sized Vehicles** between 3,500 kg and 7,500 kg but with a trailer of more than 750 kg. The weight of the vehicle and trailer cannot exceed 12,000 kg. The trailer must not be more than the maximum permissible towing weight for that vehicle.

Drivers may also be subject to additional restrictions on their licence. For further information contact the DVLA for information on drivers' licences and ask for booklet INS 57 P. This booklet can also be downloaded off the Internet.

Volunteer Drivers

Volunteers who are helping organisations such as Community Groups, Scouts and Guides may drive a vehicle with only the Category 'B' licence provided they operate under the following rules:

1. They are 21 years of age or older
2. They have held the Category 'B' full licence for two or more years
3. They are providing that service on a voluntary basis (unpaid)
4. The permissible maximum weight of the vehicle is usually no more than 3.5 tonnes, excluding any specialist equipment for the carriage of disabled persons. An allowance of 750 kg is given for this equipment including items such as wheelchairs. This takes the maximum permissible weight to 4.25 tonnes
5. They have received training and an assessment in that type of vehicle
6. They may not tow a trailer
7. They may not drive abroad (UK only).

Please Note:

To drive a vehicle in excess of 3.5 tonnes without specialist equipment for the disabled, the driver will require a Category 'C1' or 'C' licence for large vans or lorries; or a Category 'D' or 'D1' licence for buses or minibuses.

Fitness to Drive and the Importance of Well Being

As every driver knows, to drive a large vehicle you need to be relatively fit. However, for professional drivers, regular health checks and assessments every five years 'maximum' are part of the job. While casual drivers do not need to undergo regular medicals, they do need to ensure that they are in good health for the safety of themselves and their passengers as well as that of other road users.

Eyesight and Vision

To be able to drive any vehicle, there is a minimum standard of sight required. You must be able to read a number plate at a distance of 20 metres or 65 ft. If you need glasses or contact lenses to read the number plate at that distance, you must use them when driving. The police have the power to make a driver undergo an eyesight test.

Driving after drinking alcohol or taking drugs

According to Dft figures, one in seven of all deaths on Britain's roads is alcohol-related. This means that in the UK, about five people are killed each week by drunk drivers.

Despite high-profile government campaigns targeting drivers, and ever harsher penalties imposed on those who drink alcohol before driving, thousands of people continue to flout the law, putting their own lives and those of others, at risk. It is well known that alcohol depresses the body's physical and mental responses and our ability to make reasoned and sensible judgements. Anyone who has ever consumed a large amount of alcohol will tell you of their inability to reason, think or undertake a range of quite simple activities. Driving any vehicle after drinking alcohol, even if you are an experienced driver, will place you and other road users at risk.

Drink-driving: limits and penalties

So how much alcohol can you consume before you are over the legal limit? The answer is that we simply don't really know. There are so many variables: the weight and physical condition of the person; if they have had a meal; if they are taking any medication. Many other factors can also affect their condition.

The UK drink-drive limit when measured is:

- Breath is 35 micrograms of alcohol per 100 millilitres of breath
- Blood is 80 milligrams of alcohol per 100 millilitres of blood
- Urine is 107 milligrams of alcohol per 100 millilitres urine.

If a driver is found to be above the legal limit, they can be jailed for up to six months, fined up to £5,000 and disqualified from driving . In most cases, disqualification is automatic.

You should be aware that, even if you are just marginally over the limit, you will still probably lose your licence for 12 months. If it is a second offence within ten years, you could receive a three year ban.

In the far more serious case of causing death while driving under the influence of alcohol, the offender can receive a maximum sentence of 14 years in prison, an unlimited fine and an automatic two year ban.

Many drivers fail to understand how long alcohol can stay in the body and have found, to their cost, that they have failed a breath test when driving the following morning after having had an evening out.

Drugs and Driving

When people hear or see the word 'drugs', they generally think of illicit substances. The truth is that many of us take some form of drug/medication when we are ill or have a medical problem. In some cases, these can be as dangerous as alcohol. Drivers need to be aware that there are also many drugs, available over the counter without a prescription, which can affect your ability to drive.
Some of the more common medicines can make you sleepy and impact on your ability to concentrate.
If you plan to take medication for anything, read the label first!

Smoking

With changes in the law, it is now illegal to smoke when on board your vehicle, even if you are not driving. This is because it is classed as your 'place of work'. Drivers who smoke while in their vehicle can attract a fine of up to £200.
Smoking can also damage your health. A recent survey shows that people are now living longer. One reason for this is that the number of people smoking is reducing. This increase in longevity (the number of years we live) is set to increase. The average life expectancy for a man is now 80 years and for a woman, 83 years, (although men are catching women up). However, the average life expectancy for smokers is significantly less at around 65 years. So, do the maths - if you smoke you could be giving away years of your life.

Stress

It is no secret that driving aggressively and reacting to other drivers on the road, increases your stress levels. Government figures show that up to 60% of all absences from work are caused by stress.

We know that stress can make us depressed and angry with others. When driving, we need to ensure that we stay calm, and in control, at all times.

It is not easy when you are on a timetable but try and allow extra time on the road. Cutting it too close from the start of your journey, can increase blood pressure and cause a lot of stress because you feel that you need to drive faster and try and dodge traffic. It also increases the chances of being involved in road accidents. If you can, leave early, as it gives you more time to relax a little on the road. Where possible, it also helps not to drive during peak times.

Road Rage

The UK has one of the worst records in the world for road range incidents, with men being more likely to commit an aggressive act than women.

Some people feel that when sitting in their cars they can behave badly because they have anonymity (nobody knows who they are). This is certainly not the same for you if your are driving a large, sign written vehicle. The fact is that motoring is about dealing with other people and, most importantly, yourself. If you lose self-control you have lost the argument. Drivers must be self-aware. Dealing with, and reacting to, bad driving, is another form of conflict that can be a major cause of road traffic accidents.

The following strategies will help you to deal with a possible road rage situation:

- Don't look for conflict
- If someone is being aggressive, keep calm
- Avoid physical contact
- Take a non-aggressive posture - don't invade their space

- If you speak to them, ask how you can help with any problem they may have
- Try not to be negative
- If possible, try and inject some humour into the situation
- Control your breathing
- Speak softly and don't raise your voice
- If all else fails, walk away.

Driver Fatigue

Driver fatigue is a very dangerous condition, often resulting in an induced hypnotic state. This effect, especially during night time, can result in drivers either falling asleep at the wheel or being so exhausted that they make serious, and sometimes fatal, driving errors.

The early hours of the morning and the middle of the afternoon are peak times for fatigue accidents. Also, long journeys on monotonous roads, such as motorways, are the most likely to result in a driver falling asleep.

Sunlight signals our bodies when to be awake, but when deprived of any natural light at night, we will feel a surge of fatigue. In the middle of the afternoon, we feel fatigue to a lesser extent, particularly after a heavy meal or alcohol. The latest research shows that the grogginess felt right after you wake up can also be dangerous.

Figures from the Dft show that 300 people a year die from sleep-related road accidents in the UK.

Understanding your vehicle and its characteristics

Before you drive your vehicle, it is vital that you have a complete understanding of the vehicle, its characteristics and how it will perform. This is of even greater importance if it is your first time driving that particular vehicle.

Safety is your main priority, so get to know the vehicle you are going to drive. Familiarise yourself with all the controls and the seating position, making any adjustments before you move off. Use your walk around checks to get to know and understand your vehicle better.

Take a good, hard look at your vehicle, getting to know its physical dimensions, size and weight. You will need to know and understand the following:

- Its **length** when cornering and overtaking
- Its **height** when driving under bridges, trees and other possible obstructions
- Its **width** when passing through narrow gaps and openings
- Its **weight**, both loaded and unloaded.

As a driver, it is your responsibility to understand your vehicle, its characteristics and performance. You will need to know and understand the following:

- How long it will take for your vehicle to stop when loaded or unloaded.
- How to control your vehicle at all times, particularly when travelling downhill.
- How much time and distance you will need to overtake.
- How much power and which gear to use when travelling uphill.
- How to control your vehicle and make subtle and not violent changes to the controls.
- How to attach and detach trailers safely (if required).
- How your vehicle behaves when loaded and unloaded.
- How to load and unload your vehicle safely.
- How to distribute and secure any load (luggage).
- How to operate all auxiliary equipment on the vehicle.
- How your vehicle will behave in poor weather conditions such as high winds.

You need to understand your vehicle and all its idiosyncrasies and characteristics. If it is a new vehicle or not your usual vehicle, you will need to take time to get to know and understand how it is going to behave under all driving conditions.

Vehicle safety and daily walk-around checks

The safe operation of your vehicle is vital, not only to you but also to the public at large. Therefore, drivers need to undertake daily checks of their vehicles in order to ensure that safety is a priority. Use a checklist to help you with this task.

It is the responsibility of the driver to check the vehicle before driving it. All vehicles should be checked whether the vehicle is owned, borrowed or hired.

Should you find a problem, you should report it to your supervisor, employer or maintenance team. Take a good look at the reporting documents and familiarise yourself with them.

When reporting faults, you should use a bit of common sense. First, all faults need to be reported in writing. This helps to ensure that faults get put right and not forgotten, which can happen even with the best of employers. If it is a more serious fault which affects the safety of the vehicle, report it immediately. More minor faults found during the day can be reported at the end of your working day.

You should also be aware that if you don't carry out a check on your vehicle and it has had minor accident damage caused by the previous driver, you could end up getting the blame. As a driver, you are responsible for your vehicle.

In addition, you need to be aware that the defect report form is a legal document and once the fault has been rectified by the engineer/mechanic, they should 'sign it off'.

All maintenance records should be kept by the operator and made available for inspection by the DVSA for 15 months from the date on that record, unless it is a nil report, in which case it is usually kept until the vehicle's next safety inspection.

Remember, as the driver, you are in control of your vehicle and safety is your top priority!

Passenger Assistants

Ideally, there should be two adults on the vehicle, particularly if carrying disruptive or vulnerable passengers. This allows the driver to focus on the driving while the second adult can deal with any passenger issues.

Passengers with Special Needs

When carrying passengers with special needs, the driver and passenger assistant need to be aware of the wide range of abilities and any disabilities each passenger may have. Special needs can be physical, mental or emotional and in many cases, it may be all three. Drivers and passenger assistants should be knowledgeable and familiar with the needs of each individual in their care to be able to give them the specific support they require.

Passengers in wheelchairs will require the use of specialist equipment such as ramps, lifts and anchoring systems for the wheelchairs. Drivers and passenger assistants will need formal training in the safe use of this equipment to reduce the risk of injury to all. In addition they should know how to locate safely and secure the wheelchair and passenger without causing undue stress or anxiety to the passengers. The driver and passenger assistant should not attempt to use such equipment until training has been completed. Formal training will also be required for dealing with emergencies such as passengers becoming unwell or accidents.

It should be noted that there is a legal requirement to ensure that all lifting equipment (including tail lifts) are inspected at regular intervals in accordance with the guidance set out by the certifying body.

The Blue Badge Scheme

This scheme is designed to enable badge holders to park their vehicles closer to the desired location than would normally be permitted. This does not mean that vehicles can be parked anywhere and restrictions still apply. Drivers should inform themselves by reading the written advice given to all those who are in receipt of the badge. The advice includes information on where and how the badge should be used.

Seatbelts and the law

In general, the law states that if a seatbelt is fitted, it must be used. However, in practice, it is more complicated and all drivers, particularly coach and minibus drivers, need to be aware of the regulations.

In **1983** the wearing of seatbelts in cars became compulsory and in **2006** **f**urther regulation was introduced covering the wearing of seatbelts in cars, small minibuses under 2,540 kg, light vans, medium and heavy vehicles. See **Seat Belt Chart** below.

Vehicle occupant	Front Seat occupant	Rear seat occupant	Who is responsible?
Driver	Seatbelt must be worn if fitted	Not applicable	Driver
Passengers 14 years and older	Seatbelt must be worn if fitted	Seatbelt must be worn if fitted	Passenger
Children up to 1.3m or approx. 4ft 5ins tall = 12 to 13 years old	The correct child restraint must be used	If a seatbelt is fitted, the correct child restraint must be used	Driver
Children up to three years old	The correct child restraint must be used	If a seatbelt is fitted, the correct child restraint must be used	Driver
Children more than 1.3m or approx. 4ft 5ins tall = 12 to 13 years old until 14th birthday	Seatbelt must be worn if fitted	Seatbelt must be worn if fitted	Driver
Occupants of buses and coaches including minibuses over 2,540 kg			
Seated passengers over 14 years of age	Seatbelt must be worn if fitted	Seatbelt must be worn if fitted	Passenger
Children aged three to 14th birthday	Seatbelt must be worn if fitted	There is no legal requirement for passengers to wear a seatbelt	Driver

Seat Belt Exemptions
It should be noted that some drivers and some occupants are exempted in some vehicles from wearing a seat belt. These may include drivers of local delivery vehicles such as postal vans, milk round delivery vehicles, as well as taxi drivers whilst driving locally.

Exemption Certificates
It is possible to obtain an exemption certificate from your doctor allowing you to travel in a motor vehicle without having to wear a seatbelt. In such cases, the certificate must be carried with you each time you travel. Failure to carry the certificate may result in a fine if stopped by the police. In addition, it is advised that any passenger who is not wearing a seatbelt should not be seated behind a passenger who is wearing one. This is in order to prevent possible injury to the seatbelt wearing passenger by the unrestrained passenger in the event of a collision.

Penalty for not wearing a seatbelt
The current penalty for not wearing a seatbelt and not complying with the regulations, is a maximum fine of £500. However, you will not be awarded points on your licence and it is more usual to receive a fixed penalty notice of £60. That said, the cost of not wearing a seatbelt, if involved in an accident, could be significantly greater.

Cartoon by Master Sgt. W.C. Pope

SPEED LIMITS

Type of Road	Built-up Area	Single Carriageway	Dual Carriageway	Motorway
Types of vehicles	Speed in mph (km/h)	Speed in mph (km/h)	Speed in mph (km/h)	Speed in mph (km/h)
Car and Motorcycles (including car derived vans up to two tonnes maximum laden weight	30 (48)	60 (96)	70 (112)	70 (112)
Car towing caravans and trailers including car derived vans	30 (48)	50 (80)	60 (96)	60 (96)
Buses, coaches and minibuses (not exceeding 12 metres in overall length)	30 (48)	50 (80)	60 (96)	70 (112)
Goods Vehicles (not exceeding 7.5 tonnes maximum laden weight)	30 (48)	50 (80)	60 (96)	70 (112)
Goods Vehicles (exceeding 7.5 tonnes maximum laden weight)	30 (48)	50 (80)	60 (96)	60 (96)

Drivers hours and rest periods

It is vital that drivers do not drive when tired. Indeed, some studies have shown that driving while tired is as bad as driving while under the influence of alcohol or drugs. Professional drivers are restricted on the number of hours they may drive before they must take a break and the number of hours they may work in any one day, week or fortnight. (see chart below)

Rest Periods	As a minimum, drivers must take an uninterrupted break of 45 minutes after each 4.5 hours of driving.
Daily Driving	Drivers should not drive for more than a maximum of nine hours in any one day. However, this can be extended by one hour to ten hours twice a week.
Daily Rest	Drivers should take 11 hours daily rest. However, this may be reduced to nine hours under certain circumstances.
Weekly Rest	Drivers should take a break of 45 hours after six days of working. However, again, this may be reduced under certain circumstances.
Fortnightly Driving	Drivers may not drive for more than 90 hours in any rolling fortnightly period.

In addition to the rule on drivers hours, it should also be noted that drivers are also subject the EU rules on working time.

While the above refers to driving time for professional drivers, many drivers may find themselves driving after a hard day undertaking other duties e.g., a teacher may drive a minibus on a long trip after teaching all day.

So before driving, they (and their managers) need to consider how they (the driver) may feel when driving back at the end of the day. How safe would it be for them to undertake that trip? Both the driver and manager have a 'duty of care' to all.

Casual or non-professional drivers should also restrict the numbers of hours they drive in order that they are not too tired or lose focus. According to the Department for Transport (Dft) falling asleep at the wheel is one of the biggest causes of single vehicle accidents.

Passenger safety and comfort

In order to ensure that your passengers are safe throughout their journey, you will need to have an understanding and knowledge of the following:

- Ensuring the safety and comfort of your passengers throughout their journey.
- Understanding the forces affecting your vehicle and its passengers.
- Managing your passengers and conflict.
- Seatbelt regulations.
- The carriage of people with disabilities or special needs.
- Understanding the regulations governing the carriage of passengers in the UK.
- Understanding the regulations governing the carriage of passengers internationally if travelling overseas.

Ensuring the Safety and Comfort of Passengers

As a professional or casual driver, your first priority is to ensure the safety and comfort of your passengers throughout their journey. So what do you need to do to achieve this goal?

The key to all that you do and achieve is attitude. It sounds simple but having a positive attitude to your work and passengers will shine through

What practical steps can you take to enable your passengers to have a safe, comfortable and trouble-free journey?

1. Make sure you are fit to drive.
2. Complete your daily checks and ensure your vehicle is safe to drive.
3. Plan your journey.
4. Load all the luggage safely, ensuring that all emergency exits and walkways are clear with no overloading of the vehicle or axles.
5. Ensure the safety of others.
6. Be in control of yourself and your vehicle at all times.
7. Be patient with others and drive with care.
8. Be observant and try to anticipate events.

1. Fitness to drive

You must ensure that, before you start your journey, you are fit to drive your vehicle. Make sure that:

- You have rested and had enough sleep.
- You feel well enough to drive.
- You are relaxed and not stressed.
- You have no major ailments which could affect your ability to drive.
- You are not on any medication which could affect your driving.
- You have not consumed alcohol or taken any prescription or over the counter drugs which could affect your driving ability.

2. Daily Checks

Completing your daily check list and ensuring that any faults are attended to, is key to ensuring that you have a trouble-free journey and that your vehicle is safe to drive. The daily check list is covered elsewhere in this booklet. You employer should provide you with a daily check list in order that nothing is left out or forgotten during your check.

3. Planning your journey

When planning your journey, make sure of the following:

- That you know where you are going.
- That you are satisfied you have chosen the best possible route.
- That you have checked for traffic hold-ups, as far as possible.

- That you have planned your breaks and stops.
- That all drivers are following the same route, if travelling with other vehicles.
- That you plan to use dedicated bus lanes in traffic to save time.

4. Loading luggage safely

It is important that you load your vehicle with care to ensure the following:

- You don't overload it by exceeding its maximum permissible weight.
- You don't exceed the maximum permissible weight for each axle.
- If making a number of stops, ensure, as far as possible, that the luggage is stowed in such a way as to give access to luggage at each stop and is not blocking any exits or walkways.
- Encourage passengers to sit spread out around the vehicle and not all on one side.
- Ensure wheelchairs and their users are fully secure in the vehicle.
- That emergency exits and walkways are clear of any obstructions.

5. Ensuring the safety of others

Ensuring the safety of others is a top priority. This includes other road users as well as any passengers and colleagues. You should ensure the following:

- Your passengers are seated before moving off.
- Where possible, discourage your passengers from leaving their seats while the vehicle is moving.
- Encourage adults to wear their seatbelts and advise them that it is the law. (Those under 14 years of age on the school run may need a firmer prompt).
- Always drive with care, taking account of road conditions.
- Always be aware of pedestrians and other road users.

6. Being in Control

Being in control of yourself and your vehicle at all times, is essential for the safety and comfort of everyone on board. You should :

- Do not react to aggression from others.

- Avoid behaving aggressively with, or towards,other road users with your vehicle.
- Do not let yourself be goaded into an aggressive reaction when others are aggressive towards you. Instead, just back away.
- Do not behave in competitive manner with other road users by racing your vehicle.
- Never take risks or show off with your vehicle.
- Avoid making sudden movements with your vehicle. These include acceleration, braking and cornering.

7. Being patient and driving with care

Being patient with others (passengers, colleagues and other road users) allows you to drive with care. You need to remember at all times that safety is your main priority and you should:

- Allow for other people's mistakes.
- Consider others before yourself.
- Always drive to the best of your ability.
- Be grateful that other people's failings are not yours.
- Be content. You don't have to prove yourself by driving competitively.

8. Be observant and try to anticipate events

Being observant and anticipating events can prevent you getting into trouble. Develop these skills by:

- Planning ahead for how you will deal with the next road hazard.
- Remembering many other road users may not have the same high level of skills as you.
- Always consider the size of your vehicle and the position of vulnerable road users.
- Anticipate that other drivers may do the unexpected.
- Adopt a safety-first approach to everything you do.

Understanding Risk When Driving

When we drive, it is vital that we assess the risks to ourselves and others. If we failed to do this, we would find ourselves involved in incidents, of one form or another, every day. (In the UK, every day, five people die on our roads. That's over 1750 people a year. Thousands more are injured - many with life-changing injuries.)

Drivers need continually to ask themselves questions such as:

- Am I driving too fast for this type of road or for the conditions?
- Am I too close to the vehicle in front?
- Could there be a hazard around the next bend, such as a broken down vehicle?

It is a well-known fact that the drivers who question themselves and their abilities, are the ones with the best safety record.

Passenger Assistants

If, as a driver, you are fortunate enough to have a passenger assistant on your vehicle, you must make full use of their knowledge and skills. They are there to support you as a driver, by promoting and maintaining good behaviour from the passengers being transported.

Both driver and passenger assistant need to formally risk assess and identify hazards that may arise from the needs of some of the passengers on the vehicle.

The driver and passenger assistant should be fully aware of each other's role in ensuring that the safety of all is the priority. A good working relationship is key to ensuring that this is indeed the case.

Briefing the passenger assistants

Drivers and passenger assistants should be friendly and welcoming towards those in their care. Passengers will need reminding of the rules when travelling on the bus as well as being made aware of the following:

- That all those travelling on the vehicle must wear a seat belt.
- The standards and behaviour expected of all the passengers.
- The location of the emergency exits and when they should be used.
- What they need to do in an emergency.

Managing your Passengers and Avoiding Conflict

Drivers may experience rowdy, aggressive and even violent behaviour from passengers. Dealing with aggression and conflict is not easy and can be quite a stressful situation. It is worth reiterating here that you are responsible for your vehicle and the safety of all your passengers. You don't want to take any action which may endanger yourself or others. Some passengers may have behavioural issues which may require specialist individual support.

So, how would you deal with an individual intent on behaving badly and therefore causing a great deal of stress to you and other passengers?

Don't look for conflict. If someone has spoken to you in an aggressive manner, do not respond in a similar way. If you do, it can escalate an already tricky situation. Try to behave professionally at all times and do the following:

- Stop your vehicle in a safe place to allow you to help sort out any problems.
- Stay calm and control your breathing.
- Adopt a non-aggressive posture by not invading the other person's space and at the same time keep your hands low with your palms open (no fists).
- Speak softly and do not raise your voice. Shouting will not help to calm the situation.
- Avoid physical contact. The last thing you should do is grab or hit the person causing the problems. If they make a complaint or are injured, you could find yourself being charged with assault or losing your job.
- Listen to what they have to say and keep eye contact.
- Try to placate them, being positive with your responses.
- Don't get excited. Continue to remain calm.
- If all else fails, back away.

Report all incidents to the transport office on your return to the depot.

Somerset County Council has given the following advice to all operators and drivers carrying pupils to and from Somerset schools.

1. Make every effort to defuse situations and avoid confrontation.
2. The safety of the passengers and other members of the public must be given the highest priority.
3. Hands off! Drivers must not touch schoolchildren in any circumstances other than genuine self-defence, medical need or the prevention of a serious offence or threat to safety.
4. Schoolchildren may not be told to get off the bus, or refused entry.
5. No racist or other offensive or abusive language will be tolerated.
6. No threats should be made but warnings can be given. Do not issue warnings unless you intend to act on them.
7. Do not react to bell-ringing or mild verbal abuse. Don't tolerate swearing or physical abuse.
8. If the bus is in motion and a warning light/buzzer indicates that an emergency exit has been opened, you should immediately stop. Do not drive with the buzzer/light sounding/showing.
9. If there is a risk of passenger causing damage to the vehicle or endangering themselves or other passengers, you should bring the bus to a halt leaving the doors open and ask them to calm down. If no response, say the following:

    I have stopped the bus because you are causing a disturbance which makes it unsafe for this journey to continue. I will remain stationary until you have all sat down in your seats and become quiet. I am about to notify your school and the Passenger Transport unit.

    If you do not return to your seats and remain quiet it may be necessary for me to call the police. In the meantime I would advise you to remain on the bus because we have not reached our destination.

    The doors of the bus are open but I would strongly advise you not to leave for your own safety.

Use of Mobile Phones and Delayed Reaction Time

Research undertaken by the Transport Research Laboratory in 2014, shows clearly how using a mobile device such as a phone when driving can endanger lives by causing a delay in the drivers' reaction time.

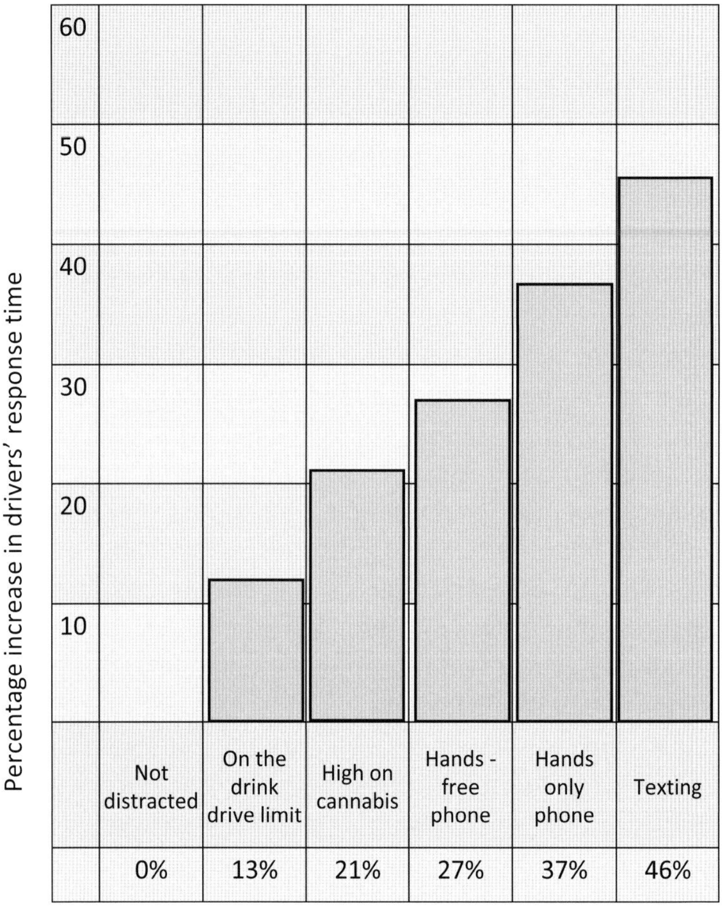

Source: Transport Research Laboratory (TRL)

Typical driver reaction time for most drivers who are not distracted or under the influence of drugs or alcohol is 1 second (shown here as 0%).

TRL's research found that the number of drivers seen using mobiles had doubled in the 3 years between 2009 and 2012. Research undertaken by the RAC also found that almost 50% of drivers aged between 18 and 24 admitted using a mobile whilst driving.
The Department for Transport (Dft) has produced data from 2012 which shows that there were 378 road traffic incidents in which mobile phones were being used these incidents resulted in 548 casualties including 17 deaths.
However, some experts believe that the figure is much higher and that in-vehicle distractions led to 9,012 incidents and 196 deaths.
In the UK, using a mobile while driving has been an offence for some time. In August 2013, the penalties were increased. A fixed penalty notice is now three penalty points and a £100 fine.
If a driver is convicted in a court of law, the fine can be increased up to £1,000 and three penalty points.
Some road safety groups are calling for a year long ban if caught using a mobile while driving. Organisations such as RoSPA and the AA are campaigning for the government to increase the penalties for using an electric mobile device while driving.

Vehicle Control

As the driver you should be in control of your vehicle at all times and in all weather conditions. In order to do this you must be aware of and take account of the following:

1. The speed of your vehicle. (The speed limit is not a target.)
2. The weather conditions (wind, rain etc).
3. The road surface (including wet leaves, camber, wet, snow and ice). Indeed, anything that may affect traction between the tyre and the road.
4. The size and weight of your vehicle and any trailer.
5. Other vehicles around you and their proximity to you.
6. Any other hazard which may affect your vehicle.

Reporting a Traffic Incident

If you are unfortunate enough to be involved in a road traffic incident, it is important that if you are able to, you need to record as much detail as possible about the event. This is not an easy task, as many people may be in shock, suffer injury as a result of the incident, or are simply so upset that they forget to take the other party's details.

If the incident involves injury, there are actions you must take by law, even if the incident was not your fault. They are as follows:

- Report the incident to the police at the earliest opportunity or within 24 hours.
- Give the other party or parties your details. These include:
  - If you are the driver, your name
  - Details of your vehicle
  - Your contact details
  - The name of your insurance company
  - The name and contact details of the vehicle owner if not you.
- If the incident did not involve any obvious injury, there is no requirement to report the matter to the police, unless you believe a crime has been committed. However, it is important to get the details of the other parties and any witnesses to the incident.
- If you have a camera, or a mobile phone with a camera, take as many pictures as possible. These need to include:
  - The other vehicle or vehicles
  - Damage to any property or buildings
  - The road layout.

Please remember if you are on a live carriageway, watch out for other traffic. Never place yourself at risk and if possible move to a place of safety.

How to Assess and Deal with Emergency Situations

Emergency situations can happen at any time, so what could be regarded as an emergency?

- A road traffic incident involving injury.
- Someone becoming seriously unwell.
- A breakdown in a dangerous situation.
- A fire on a vehicle.

How should you behave in an emergency situation?

- Stay calm. If you panic, so will other people.
- Take control. As the professional driver, you are in charge until the emergency services arrive. If you have a fluorescent jacket or tabard, put it on.
- If your vehicle is not involved, stop it in a safe place.
- If you are first to arrive at the scene, warn other drivers and contact the emergency services, before giving help.
- Only if it is safe to do so, try to protect the scene with your vehicle.
- If your vehicle is not involved, it may be safer not to evacuate your passengers.

If your vehicle is involved, follow these procedures:

- Check for injuries and, if you need to evacuate your vehicle, ensure that you can account for everyone (no-one missing) and that they are all waiting in a safe place.
- If possible, switch off the engine on your vehicle and remove the keys.
- Do not allow anyone to smoke in case of fuel leaks.
- Alert the emergency services giving exact details of location and the problem.
- If in poor light, do not stand in a position that blocks the view of the rear lights of your vehicle.
- Check to see if there are any medical professionals on board who may help with casualties. If *you* are trained, give first aid.
- Give continual reassurance to all those you are supporting.
- If you can, delegate some of the tasks to a responsible person. This could include contacting the emergency services but make sure they report back so that you know that help is on its way.

Emergency Evacuation of Minibuses

Bus operators and their drivers have a 'duty of care' with regards to all those travelling on their vehicles. As such, they must do all they can to ensure the safety of their passengers. In all cases, drivers must think clearly and make decisions regarding the safety of their passengers. **Every situation will be different.**

Drivers should ask themselves the question, 'Would my passengers be safer on or off the vehicle?' And then **'Assess the risk'**.
The following scenario illustrates the process of risk assessment and decision making:
Imagine you have had a minor incident or breakdown on the motorway and your bus is full of small schoolchildren, elderly people or students with special needs. Keeping them on the bus may well be the safest option. Small or disabled passengers will not be able to climb over the safety barriers without help, adding to the risk of them spilling into the carriageway. However, it would be wise to move any passengers out of the back rows of the bus, if possible. The reason for this lies with statistics - as many as 30% of all motorway accidents occur on the hard shoulder.

Fire and Smoke
In an incident such as a fire, or the vehicle filling with smoke, the obvious and safest course of action is to get everybody off and well away from from the vehicle and fire, upwind if possible, and away from the road. However, before undertaking this task, the driver must quickly make a full assessment of the situation so as not to expose passengers to greater danger when exiting the vehicle.

Evacuation procedure
In order to carry out this procedure as quickly and safely as possible, you (as the professional driver) must be firm and take full control. Before starting your journey, make sure you know the location of all the emergency exits and, if you can, identify a 'responsible person' on board to support you should an emergency situation arise. This could be a passenger assistant, teacher or bus prefect but do get their agreement that, in an emergency, they will help should you need to evacuate the bus. (They may need to take temporary charge of the passengers when they first leave the vehicle.) Drivers should advise the 'responsible person' of the drill and what will be required of them.
It may be that the driver has no help on board and must take full responsibility for the evacuation on his own. If this is the case, they

must give clear instructions to passengers that, on leaving the bus, they must move away from the vehicle to a place of safety.

As a driver, what should you do?

1. Stay calm, do not panic, you are the professional in charge.
2. Try to stop your vehicle quickly and as safely as possible, in a safe place.
3. Advise the 'responsible person' on the bus that there is an emergency and get them to take control of the passengers when they leave the vehicle. Try and get that person to leave the bus first. This is in order to get everyone to place of safety away from the vehicle -30 metres/100 ft minimum, if possible up wind and away from the road.
4. Open all exits on the nearside of the bus, away from any traffic. Any exits at the rear or offside of the vehicle should be a last resort option, particularly on a motorway.
5. Give the order to evacuate the vehicle and clear instructions on how this is to be done. Raise your voice so that everyone can hear what you say. Keep repeating the instructions, as some passengers may not have been listening. Try to reassure your passengers throughout this procedure.
6. Advise all passengers to leave any bags and to leave the vehicle quickly, in an orderly manner, with those nearest the exits leaving first. Advise all passengers that they must listen to the instructions from the 'responsible person' after leaving the bus and they must move away from the vehicle up wind of the fire and if possible, towards the front away from the rear of the vehicle and away from the road.
7. Try to ensure that passengers do not panic and push others, in order to get off.
8. Stay on the bus to ensure that everyone gets off as quickly as possible in a safe and orderly way.
9. Give support to any passengers who may have difficulty . If you can, get other able bodied passengers to help where necessary. Drivers need to identify vulnerable passengers as they get on the vehicle.
10. Ensure that everybody is off the bus , before leaving the vehicle yourself.

11. Only now would you telephone the emergency services. Your priority is to get everyone off first. Take back control of the assembled passengers until the emergency services arrive. If a passenger list is available , this should be checked as soon as possible, to ensure that everybody is off, safe and if they have any injuries.
12. If passengers have injuries or are in shock, keep them warm, give support and first aid, if trained.

Passengers in wheelchairs

The issue of having passengers in wheelchairs raises additional problems. They may have accessed the vehicle via a lift, with both the chair and the person strapped in. Added to this, they may not have the capacity to free themselves.

If time allows, you may be able to evacuate the passengers as you would if there was not an emergency. However, if the vehicle is on fire, or is filling up with smoke, and endangering the lives of the passengers, you have to get them off as quickly as possible. In addition, if it's an electrical fire, the lift may not work and it would be unlikely that you would have time to operate it manually.

In order to evacuate the vehicle quickly, without the use of the lift (for reasons previously stated) it may be necessary to extract passengers manually from their wheelchairs. You should do so with as much care as possible to avoid injury, distress or pain to the wheelchair passenger. Talk to them and reassure them throughout the process, which would be a last resort. Also, you must follow manual handling safety guides to avoid injury to yourself.

As with able-bodied passengers, on exiting the vehicle, disabled passengers must be moved away from the vehicle to a place of safety.

* **Footnote**

In all cases, the first priority of the driver is the safety of the passengers and not to waste time trying to fight the fire, unless doing so would help any trapped passengers to escape.

Driving

Drivers are advised to read the Highway Code before driving any vehicle, let alone one that they may not be familiar with. Put simply, there are different rules which apply to operating and driving larger vehicles. These include speed limits, the time it takes to stop and the handling of the vehicle, particularly in poor weather conditions.

All drivers who drive vehicles belonging to local authorities and private companies are required to undertake training and assessment of their driving skills, knowledge and understanding. In addition, drivers should be subject to periodic reassessment, dependent on the employer's policy. The reasoning behind the reassessment is not that an employer is being difficult. It is now part of the employer's 'Risk Assessment', as required by the Health and Safety Executive (HSE). Drivers have a 'duty of care' to ensure they are operating in a safe and sensible way and have the skills and knowledge to drive safely in modern traffic conditions.

Use of roof racks

There are a number of key issues to consider when using a roof rack. These include the following:

1. Adding to the weight of the vehicle

By placing heavy items such as luggage on the roof of the vehicle, you can easily overload the maximum axle weight of each axle or the overall weight of the vehicle. Although the maximum weight for that vehicle and each axle is shown on the vehicle plate, it would be difficult for drivers to estimate if they were exceeding those weights without placing the vehicle on a weighbridge.

2. Adding to the height of the vehicle

Again by placing items on the roof of the vehicle, a driver may well find that he then cannot access some car parks and any low level feature may well become a hazard.

3. Raising the centre of gravity of the vehicle

The placing of heavy items on the roof of the vehicle may also affect the stability of the vehicle due to the raising of the centre of gravity. The vehicle will also be more prone to the effects of adverse weather conditions such as high winds.

It is the driver's responsibility to ensure that the load and the vehicle are safe and not a danger to passengers or other road users.

The use of trailers

It may seem like a good idea to place a trailer on the back of a minibus, to carry the extra luggage or equipment for a trip or expedition. However, drivers should consider the following factors before taking this decision.

1. Do I have a licence which allows me to tow a trailer? (See the section on drivers' licences).
2. The additional length of the vehicle. A trailer can add significant length to the vehicle which should be taken into account when manoeuvring and overtaking .
3. The additional weight of that combination of vehicle and trailer will have an effect on acceleration, braking and handling and could cause the vehicle to fishtail under certain conditions.
4. When towing a trailer, the maximum speed limit on motorways and dual carriageways is reduced. In addition, vehicles with trailers are not allowed in lane three of motorways.

Bus lanes

Generally, minibuses with between nine and sixteen passengers seats are allowed to use bus lanes. However, drivers may find that on some bus lane signs, the word 'local' may be present. This means that a local traffic order is in place and those lanes will be restricted to local service buses. Care should be taken when using a bus lane as you will be undertaking other road users who may be moving more slowly or stuck in traffic.

Reversing

When reversing a vehicle such as a minibus, as the driver you are seated almost at the very front of the vehicle, so the rear of this large vehicle is some way back. A great deal of skill and practice is involved in reversing this type of vehicle successfully. Some drivers are more confident and some more skilled at this than others. Indeed, some find the whole manoeuvre quite stressful, particularly if being watched by others, both inside and outside the vehicle.

So, before undertaking a reversing manoeuvre, consider the following.

- Do I really need to reverse or can I turn further up the road?
- Do I have enough room to reverse safely?
- Is there a responsible person nearby who could watch and guide me? If so don't let them stand directly behind the vehicle.
- Will I be able to see well enough to undertake the task?

If there is no option other than to reverse, the following rules should apply.

- Position your vehicle (line it up correctly) at the start.
- Control your vehicle and its speed.
- Keep checking in your mirrors throughout the manoeuvre.
- Keep calm, pull forward and re-adjust if necessary.

Motorway driving

When driving on motorways, it should be noted that although speeds are higher, motorways have a lower casualty rate. There are a number of reasons for this.

1. All vehicles are (or should be) travelling in the same direction with lane separation for those travelling in the opposite direction.

2. The most vulnerable road users eg pedestrians, cyclists etc. are not allowed on the motorway.

However, there will always be an element of danger when driving on a motorway. It is like any other road and poses risks.

Should you break down or have a puncture on a motorway, it is the driver's responsibility to steer the vehicle on to the 'hard shoulder', switch on the hazard warning lights and get the vehicle as far away as is practical from the 'live' carriageway. (Not too close to any barriers, as you may not be able to open the doors on the nearside).

Drivers should acquaint themselves with their organisation's policy on what to do in the event of a breakdown as well as their employer's procedures on evacuation of passengers in the event of a breakdown or puncture.

International Journeys

If you are planning to drive a minibus abroad, you must be the holder of a full Category D or D1 driver's licence. If you do not have this category on your licence, you will have to undergo a medical examination and pass a four-part driving test to obtain it.

If you have those categories on your licence, you will also have to have a Drivers Certificate in Professional Competence (CPC), which involves 35 hours of additional, periodic training. Please contact your training manager for further details and advice on this training.

You should be aware that vehicles which are operated on permits cannot be used on international journeys. In this instance, a hired vehicle with an operators licence will be needed and appropriate insurance for both cover and recovery of the vehicle if necessary.

You must also acquaint yourself with the different regulations which will apply in each of the countries you plan to visit.

First Aid

FIRST AID KIT

It is advisable to have the following items in your first aid kit:

- First Aid leaflet
- Disposable gloves (non-latex)
- Range of dressings
- Range of bandages (including triangular bandages)
- Cleansing wipes
- Plasters
- Tape
- Scissors (blunt-edged)

Buses and coaches must have a first aid kit unless they are a local service.

Basic First Aid Techniques

Knowing what to do can make the difference to a person's recovery and you could even save their life!

What to do in an emergency

First of all you must assess the situation, the scene and then the casualty, using the Primary Survey (or **DR ABC**). This is a logical sequence of actions which help you to deal effectively in an emergency where you might have casualties with injuries.

D DANGER

Look for any dangers to you or the casualty before you approach, and deal with it, if possible.

R RESPONSE

Talk to the casualty and gently squeeze his shoulders to see if he responds. If he does not respond, you can assume he is unconscious.

A AIRWAY

Open his airway by gently tilting his head back and lifting his chin.

B BREATHING

While kneeling beside the casualty, check for breathing for 10 seconds, by looking along the length of his chest to see if it rises and falls, listen for normal breathing and feel his breaths on your face.
IF HE IS BREATHING CALL 999 OR 112 FOR AN AMBULANCE.

C CIRCULATION

If he is not breathing you will have to start CPR 30 compressions followed by 2 rescue breaths. Continue with this sequence until he starts to breathe normally, the medical professionals take over or you get too tired to continue.

Quiz Questions

1. Name the licence category needed to drive a minibus weighing in excess of 3500 kg without specialist equipment…
   A. 'B'
   B. 'C'
   C. 'D1'
   D. 'C1'

2. Drivers are required to be able to read a number plate at a distance of ..
   A. 10 metres
   B. 30 metres
   C. 15 metres
   D. 20 metres

3. The drink drive alcohol limit when measured in a breath sample Is …..
   A. 30 micrograms of alcohol per 100 milligrams of breath
   B. 35 micrograms of alcohol per 100 milligrams of breath
   C. 40 micrograms of alcohol per 100 milligrams of breath
   D. 45 micrograms of alcohol per 100 milligrams of breath

4. Some over-the counter drugs can inhibit a driver's ability to drive.
   A. False
   B. True
   C. Only if taken with other medication
   D. Only if the driver is unwell

5. Using a mobile communications device (phone) when driving is not illegal.
   A. Only if it is 'hands free'
   B. Only if you are texting
   C. As you can use a mobile at any time
   D. Only if you don't have any passengers in the vehicle

6. All minibuses should have a first aid kit as part of its equipment.
   A. False
   B. Only when carrying passengers
   C. True
   D. Only if a trained first-aider is on the vehicle

7. Ensuring that the passenger assistant understands their role is the responsibility of ...
   A. The passenger assistant
   B. The driver
   C. The management
   D. All the above

8. How often and by whom should the walk round vehicle checks be undertaken on your vehicle?
   A. Weekly by the driver
   B. Once a fortnight by the driver
   C. Only a trained mechanic once a month
   D. Every day by the driver

9. Minibus drivers are exempted from wearing their seatbelts when driving.
   A. True, as they are in charge of the vehicle
   B. True, as they may have to stop and get out occasionally
   C. False, as they must wear a seatbelt if fitted in the vehicle
   D. True, as seatbelts don't really do anything

10. Smoking in the vehicle is only permitted in the following situations.
   A. When there are no passengers
   B. The boss doesn't see you
   C. If you're not driving
   D. It is illegal to smoke in the vehicle at any time.

11. An emergency situation is
A. Something which is known about in advance
B. An unforeseen event
C. Can only happen once in a lifetime
D. Can only be sorted by the manager

12. When driving, why should more time and space be allowed in weather conditions such as snow, ice and rain?
A. You can drive faster
B. You may want to stay warm in the vehicle
C. You may skid (poor traction between the tyre and the road)
D. You can have fun spinning the wheels

13. Being patient with others when driving is ...
A. Just an excuse for being late
B. A good trait to have and can prevent accidents
C. A poor excuse and can lead to accidents
D. A reason for the passengers to get annoyed

14. Drivers who speed can...
A. Save on fuel as they get to their destination faster
B. Ensure that the vehicle is in better mechanical condition as there is more time for servicing
C. Expect a bonus from the boss
D. Place themselves and others at risk

15. What should drivers do after attaching a trailer on to their vehicle?
A. Forget about it until they reach their destination
B. Load up the trailer as full as they can
C. Think about how this will affect the operation of the vehicle
D. Ensure that they put extra fuel in the vehicle

Answers to quiz questions

1. C 2. D 3. B 4. B 5. A 6. C 7. D 8 .D 9. C 10. D 11. B 12. C 13. B 14. D 15. C